ARTIFICIAL INTELLIGENCE

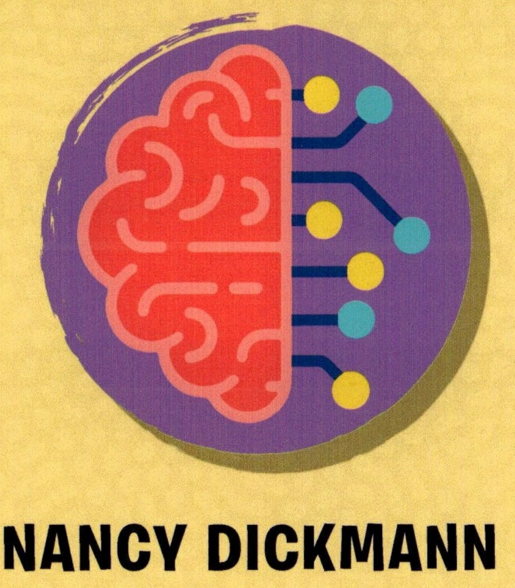

NANCY DICKMANN

BROWN BEAR BOOKS

Published by Brown Bear Books Ltd
4877 N. Circulo Bujia, Tucson, AZ 85718
USA
and
Studio G14, Regent Studios, 1 Thane Villas,
London N7 7PH, UK

Text: Nancy Dickmann
Design and Illustrations: Square and Circus
Design Manager: Keith Davis
Children's Publisher: Anne O'Daly

Library of Congress Cataloging-in-Publication Data
Names: Dickmann, Nancy, author.
Title: Artificial Intelligence / by Nancy Dickmann.
Description: Tuscon, AZ : Brown Bear Books Ltd., [2025] | Series: Science starters | Includes bibliographical references and index. | Audience: Ages 6-9 | Audience: Grades 2-3 | Summary: "Find out about artificial intelligence, the different areas of STEM behind it, and what it does"– Provided by publisher.
Identifiers: LCCN 2023052992 (print) | LCCN 2023052993 (ebook) | ISBN 9781781219874 (library binding) | ISBN 9781781219935 (paperback) | ISBN 9781781219997 (ebook)
Subjects: LCSH: Artificial intelligence–Juvenile literature.
Classification: LCC Q335.4 .D528 2025 (print) | LCC Q335.4 (ebook) | DDC 006.3–dc23/eng/20231220
LC record available at https://lccn.loc.gov/2023052992
LC ebook record available at https://lccn.loc.gov/2023052993

Websites
The website addresses in this book were valid at the time of going to press. However, it is possible that contents or addresses may change following publication of this book. No responsibility for any such changes can be accepted by the author or the publisher. Readers should be supervised when they access the Internet.

Words in **bold** appear in the Glossary on page 23.

CONTENTS

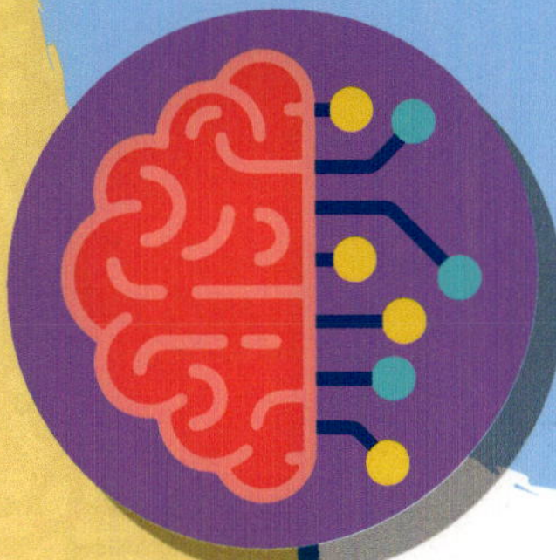

COMPUTERS THAT THINK

You probably use a computer almost every day. Computers are really useful. They're great for doing homework and playing games. It might seem like your computer is really smart. But it can only follow its **program**. It can't think like your brain does.

THE NEXT STEP

Artificial intelligence (AI) is different. It's a step up from a regular computer. You've probably heard people talking about AI. That's because it's becoming more common. New advances are being made. Now there are computers that can outperform human brains!

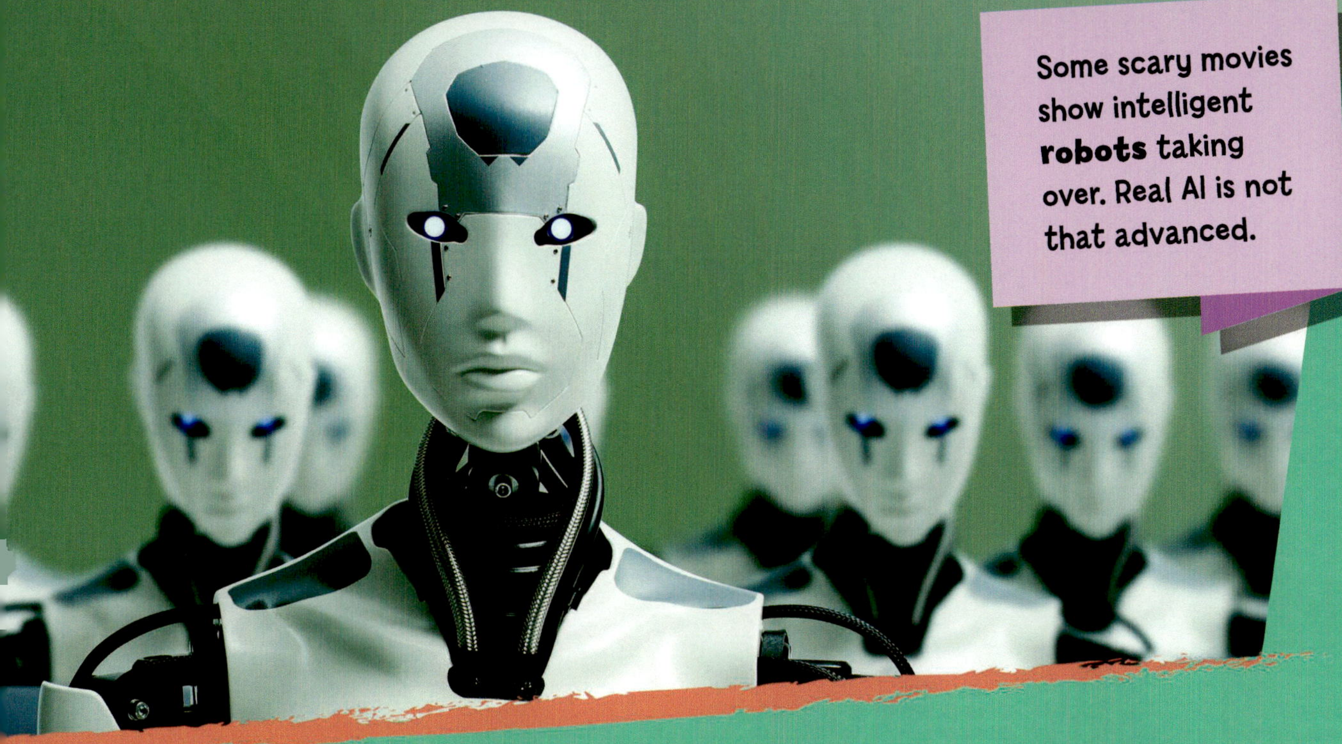

Some scary movies show intelligent **robots** taking over. Real AI is not that advanced.

Your Amazing Brain

A human brain can do a lot! Computers can do some of these things. But they can't do all of them.

Use imagination and create things.

Feel **emotions** such as love, fear, and anger.

Learn and remember facts.

Keep your body running.

Understand the outside world.

Do two things at once.

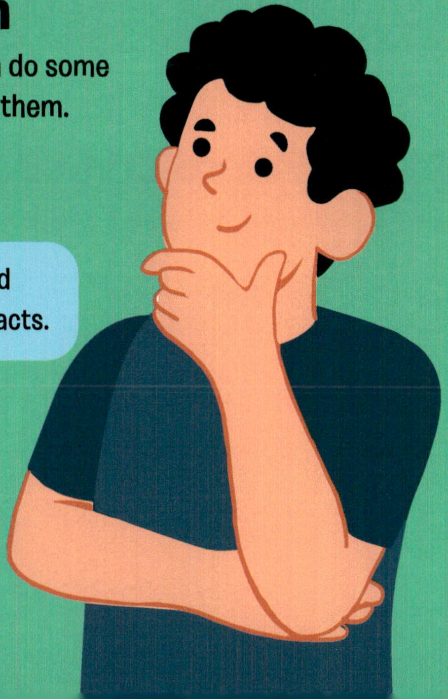

LEARNING PATTERNS

Dogs come in a range of shapes, sizes, and colors.

On your way to school, you see a dog. You've seen lots of dogs before. But this is a different breed. You haven't seen it before. It doesn't look quite the same. But you can still tell it's a dog! Your brain has recognized it. How does that happen?

MAKING CONNECTIONS

Your brain remembers other dogs you've seen. It figures out what they have in common. This forms its definition for "dog." Then it compares the new dog. Does it match? Many AIs can do something similar. They learn to spot **patterns**. They can tell if new items match the pattern.

Checking a List

What would you put in your definition of "dog"? Would these facts be enough to tell a dog from a cat or a mouse? Can you think of others to include?

A dog has four legs.

Its body is covered in fur.

Its nose sticks out like a snout.

It has a tail that it can move.

It has whiskers.

It has a short neck.

Patterns in Text

An author might use particular words or phrases a lot. An AI can be trained to spot these patterns. They can tell who wrote a piece of text.

WHAT AI CAN DO

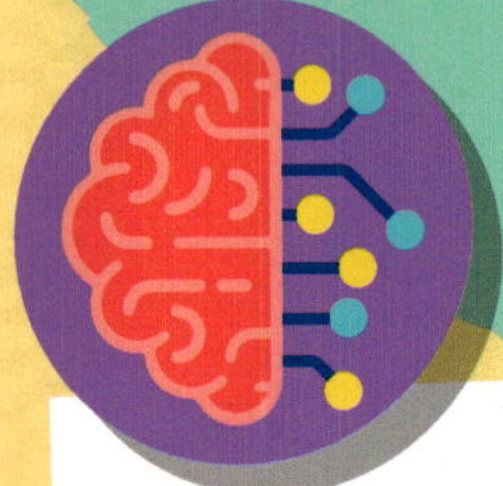

There are billions of people in the world. We each make decisions every day. What if AI could make some of them for us? Human brains are really good at lots of things. But AI is better at some. For example, it can spot patterns in huge sets of numbers.

WORKING FASTER?

Computers can sift through **data** really fast. Using AI can speed up some tasks. What if 100 people apply for a job? A person could read each application. But an AI could process them faster. It would look for the right skills and experience. This would save time.

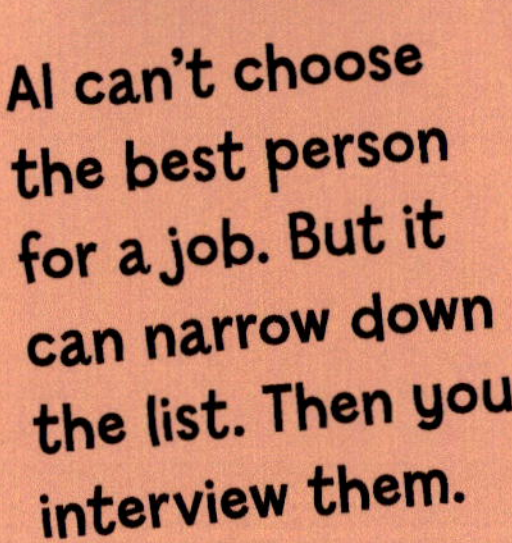

Help factory robots do their job better.

Suggest the fastest way to get somewhere.

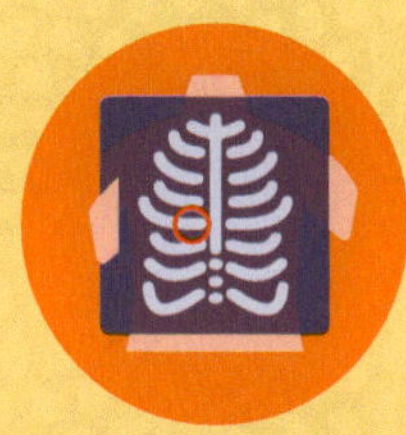

Spot medical problems on scans.

Help cars or buses drive themselves.

How We Use AI

We are already using AI in many areas. These are some of the things that AI can do.

Chat online to help customers solve problems.

Find fake reviews on online shops.

AI THAT CREATES

Some AIs can produce **content**. They are called generative AI. Some can write text. Others can produce images. They might look like paintings or photos. Some AIs can make music. Others make videos or games.

WORKING FROM EXAMPLES

A **chatbot** like ChatGPT has seen lots of text. It looks for patterns. These show it how humans put sentences and paragraphs together. It learns how reports, emails, and stories are different. Then it can write its own. AIs for images and music work in a similar way.

AI in Films

AI can make actors look younger. The actor shoots the scene. Then AI replaces their face with older footage. This is useful for flashback scenes.

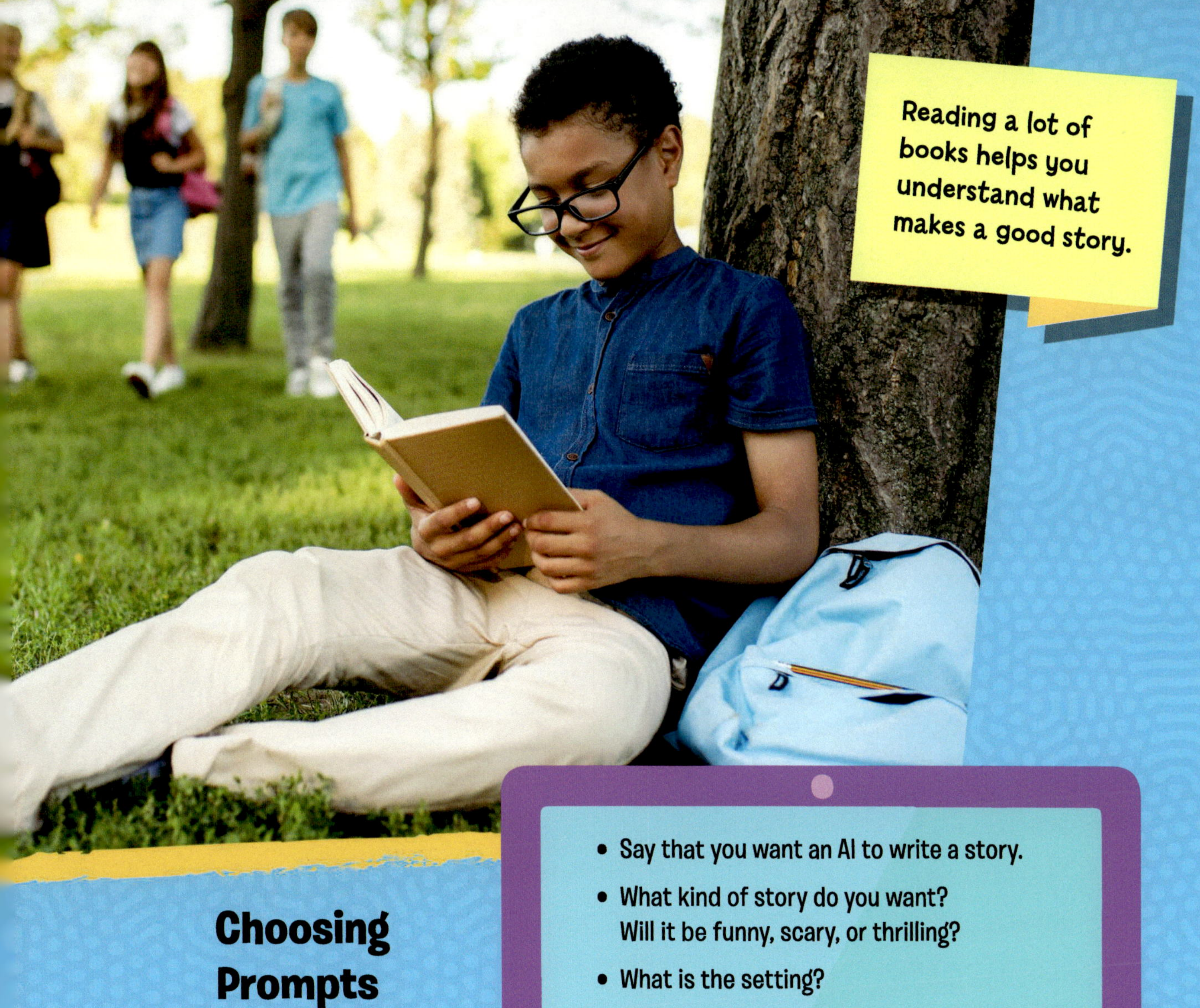

Choosing Prompts

An AI has to be prompted. That means giving it instructions. The more detailed, the better! You're more likely to get what you want.

- Say that you want an AI to write a story.

- What kind of story do you want? Will it be funny, scary, or thrilling?

- What is the setting?

- Who are the characters?

- How long should it be?

- What will happen at the end?

WHAT AI CAN'T DO

An AI can't do everything. Humans feel emotions, such as joy and sadness. An AI can give definitions for these words. But it doesn't understand the emotions. It can't form a bond with someone. An AI wouldn't make a good **therapist**.

THINKING OUTSIDE THE BOX

Some AIs are good at making pictures and text. But they don't have new ideas. Their work is based on what they've been shown. Remember, though, that AI is improving all the time. Maybe this page needs a new name! How about "What AI can't do ... *yet?*"

Passing the Test

Alan Turing was a computer pioneer. In 1950, he thought of a test. It would see if a machine could imitate human behavior. Some modern AIs are close to passing it.

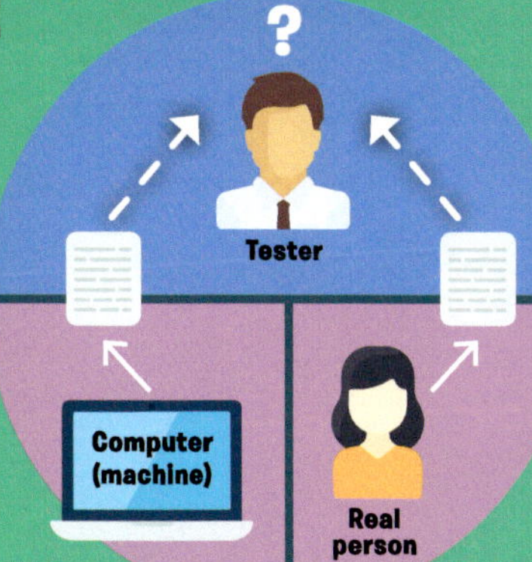

1. One person is the tester.

2. They test a real person and a computer.

3. The tester asks questions for five minutes.

4. The real person answers them. So does the computer.

5. The tester cannot see the others. The questions and answers are written down.

6. Can the tester tell which is the real person? If so, the machine fails the test.

TRAINING AND AI

How do you learn about the world? You use your **senses**. You try things out. You copy what others do. But an AI can't do this. It doesn't have eyes or ears. It's just a program. An AI needs one thing: data!

TRAIN AND TEST

You train an AI by giving it data. The AI **analyzes** the data. If finds patterns in what it's been given. Then you test it. Can it apply those patterns to new data? You tell it what it gets wrong. This helps it learn. It will do better next time.

The Right Training

An AI is only as good as the material it's trained on. Some facts on the internet are wrong. What if an AI is trained on these? It might repeat the bad information.

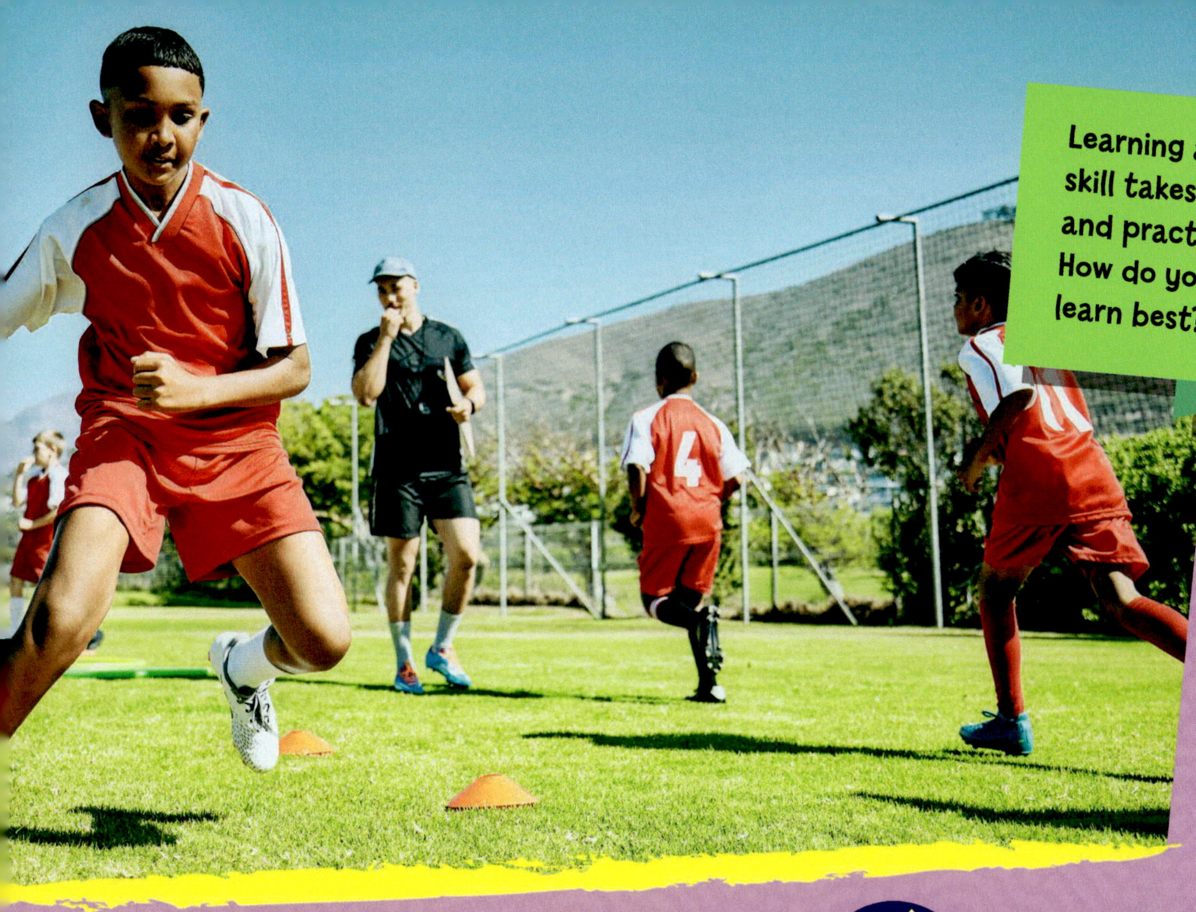

Learning a new skill takes time and practice. How do you learn best?

Training Materials

Each AI is trained for a specific job. That will decide what data it needs. Here are some of the things that AIs are trained on.

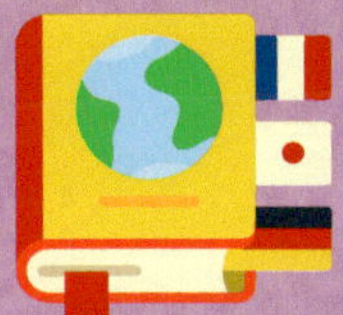

Online dictionaries and encyclopedias.

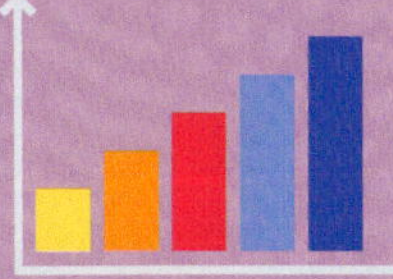

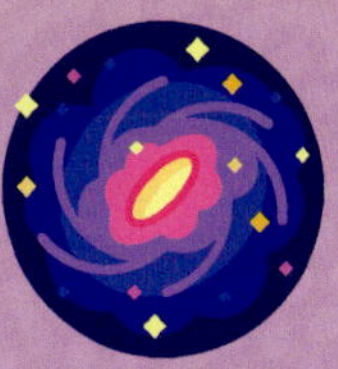

Sets of images, such as those taken by telescopes.

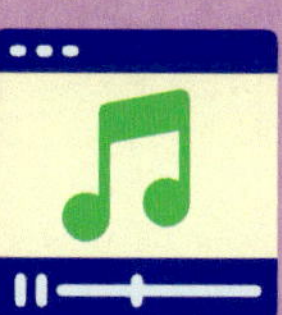

Digital music files.

Sets of numbers, such as **statistics**.

Digital books, either stories or facts.

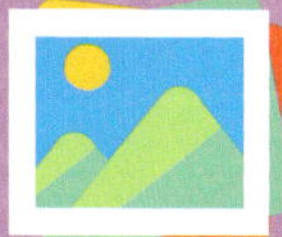

Photos with descriptions saying what they are.

AI AND ROBOTS

There are robots in movies and comics. They are in real life, too! Many of them work in factories. Some robots work in space or explore the oceans. Maybe you've built a toy robot! A robot is a machine. It's programmed to do a job. It's not the same as AI.

WHAT'S THE DIFFERENCE?

Many robots do the same job over and over. It's what they're programmed to do. But some robots are powered by AI. They can make their own decisions. They take in data. They decide the best way to get the job done. They can **adapt** if things change.

AI-Powered Robots

AI makes robots more useful. They can do more things. We are already using AI-powered robots. Here's what they can do.

Cook in a fast food restaurant.

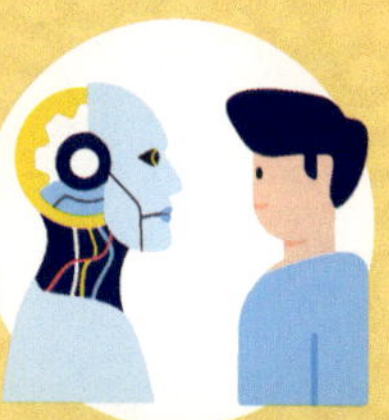

Talk and interact with people.

Walk and deliver packages.

Do construction work.

Fix machines deep underwater.

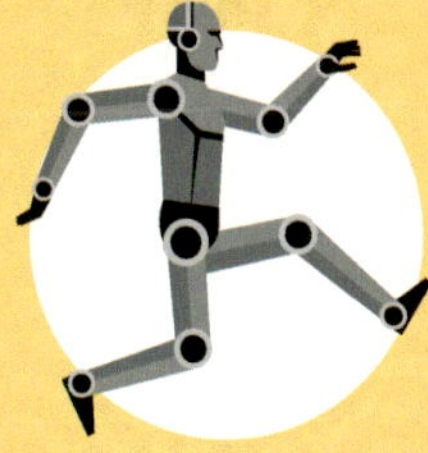

Do tricky stunts for films.

RIGHT OR WRONG?

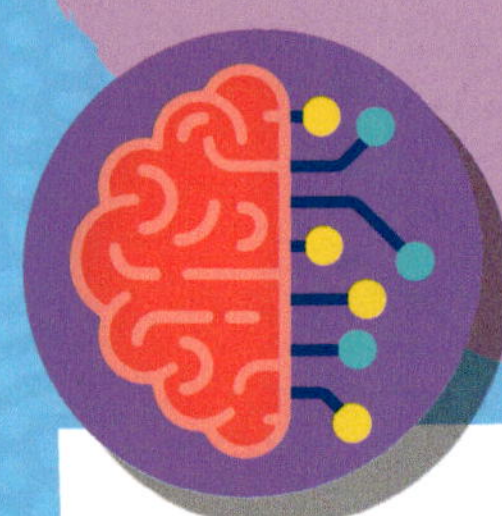

AI is pretty new. But it's getting better all the time. It will change our lives. That's why it's important to get it right. There aren't many laws about AI yet. We're still trying to figure it out. AI should be fair and safe for everyone. It shouldn't cause harm.

GETTING PERMISSION

For example, AIs are trained on data. It might be books or photos. It could be paintings or music. But what about the people who made those things? Is it fair to use their work without permission? Should they be paid for it? What do you think?

Musicians work hard to write new songs. Should an AI be allowed to copy or adapt them?

Careers in AI

AI might replace some workers. But it also creates new jobs! Many businesses want to use AI. They need people who understand how to build and train AIs.

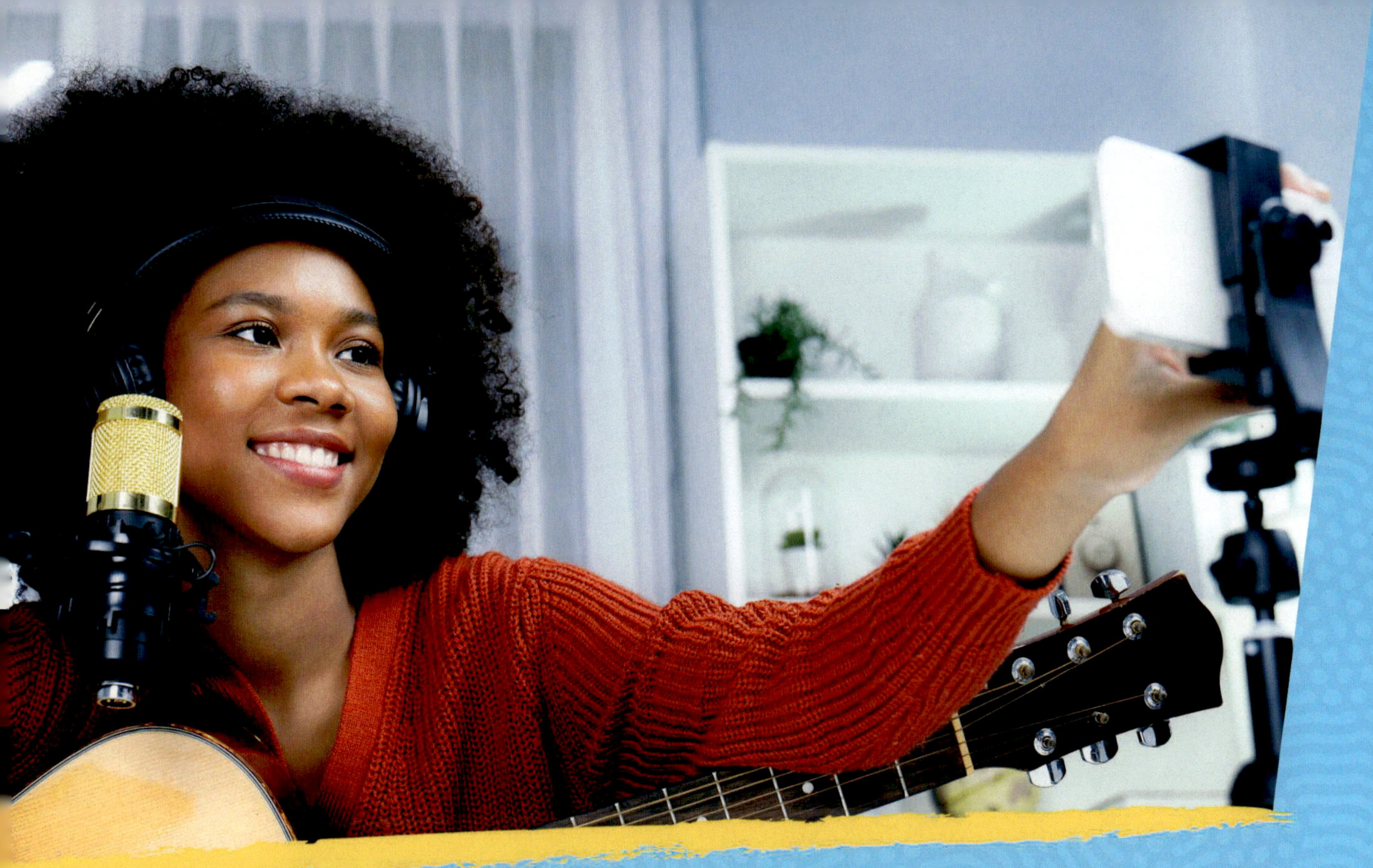

What about Jobs?

People worry that AI will take their jobs. Some kinds of jobs are more at risk. Others are safer. They're hard for an AI to do.

People who write computer programs.

Warehouse workers.

People who write sports reports.

Plumbers and electricians.

Chefs and waiters.

Doctors and nurses.

AI IS EVERYWHERE!

AI might feel like something from a sci-fi story. But it's already in our lives. You may use AI without knowing it! Do you use a streaming service for music or TV? These programs often use AI. They notice what you like. Then they suggest things that are similar.

HOW WE'RE USING AI

Some search engines use AI to show you the best results. Email programs see what you're writing. Then they suggest text for you. Doctors use AI to spot medical problems. They use it to choose the best treatment. AI really is everywhere!

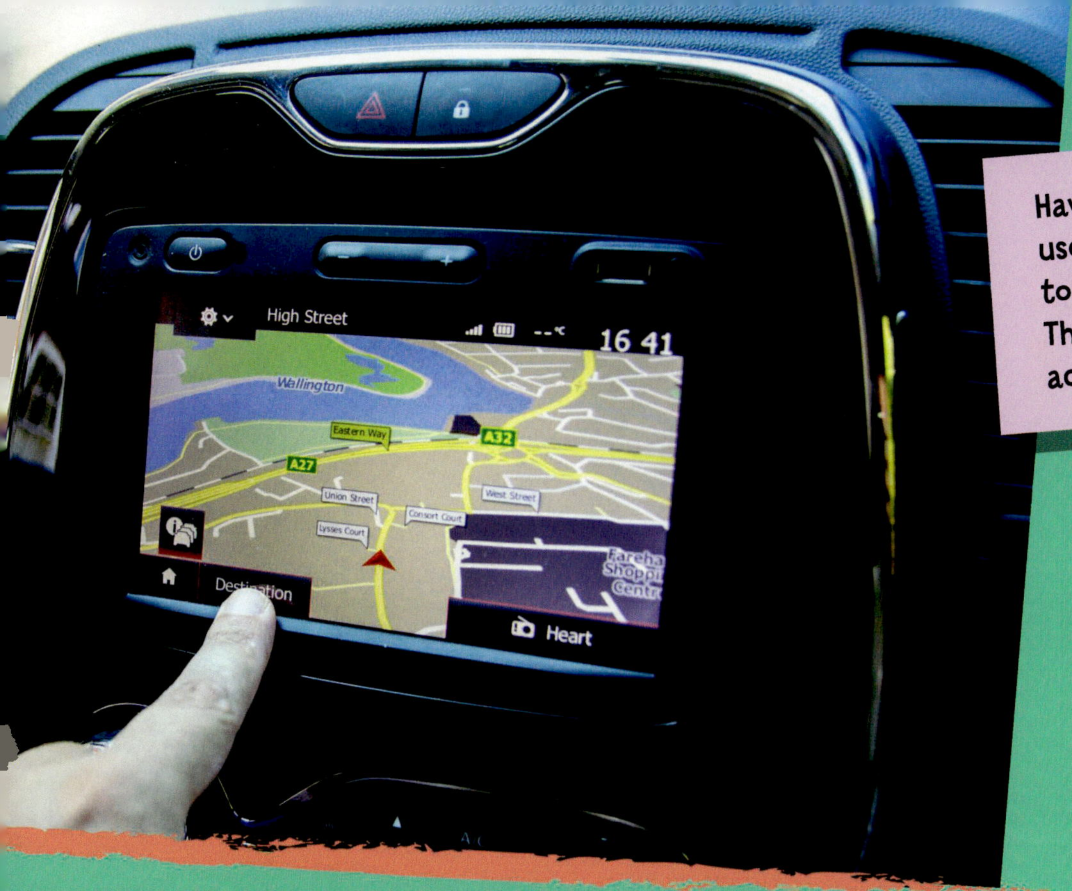

Have you ever used a map **app** to plan a route? That's AI in action.

An AI-Powered Future

In the future, we will use AI more and more. Here are some ways it might be used.

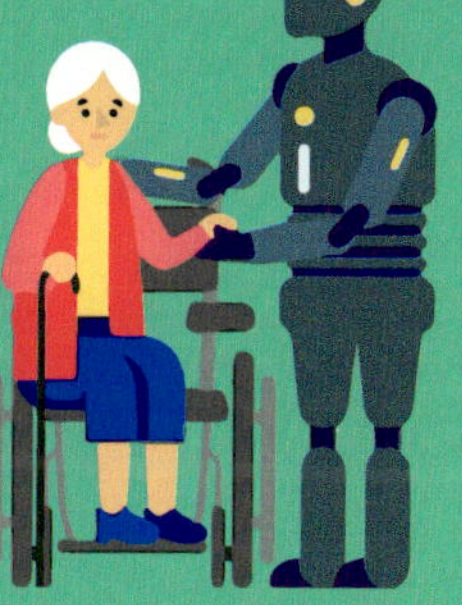

AI-powered robots take care of the elderly.

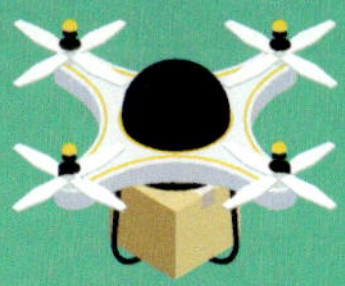

Deliveries arrive by drone.

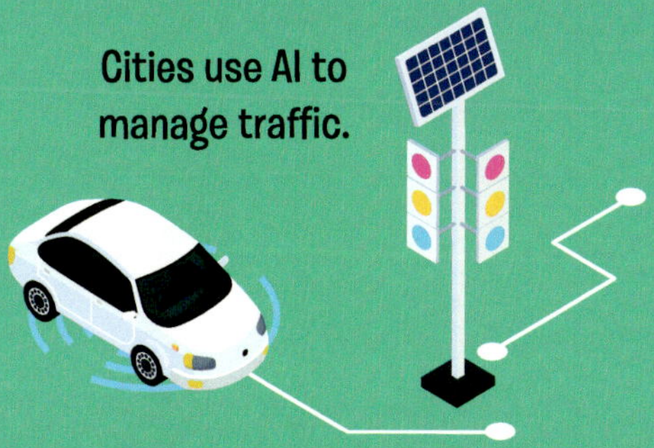

Cities use AI to manage traffic.

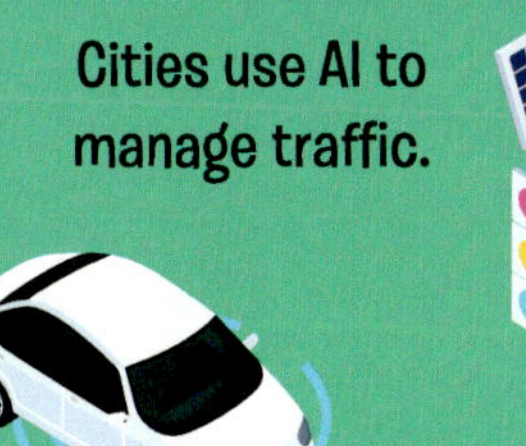

Self-driving cars carry us around.

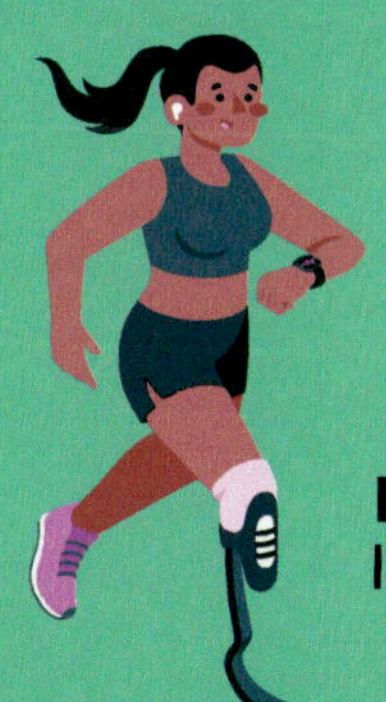

AI makes **prosthetic** limbs better to use.

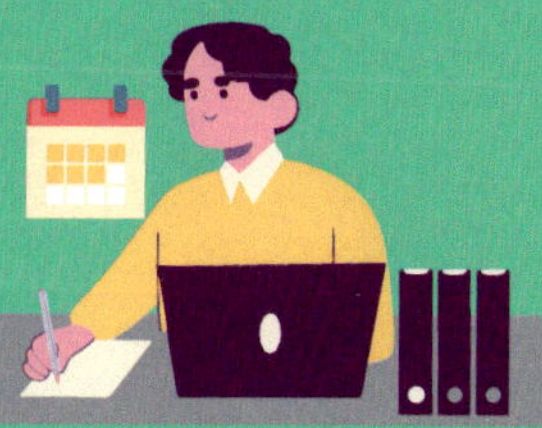

AI helps teachers track students' progress.

QUIZ

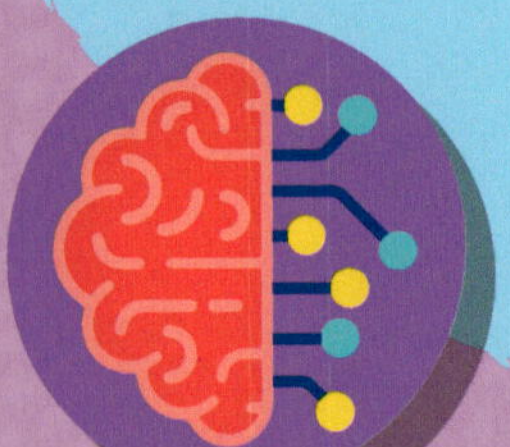

How much have you learned about artificial intelligence? It's time to test your knowledge!

1. What does generative AI do?

a. produces content such as text, images, or music

b. helps the elderly

c. powers intelligent robots

2. What do we call the instructions that you give an AI chatbot?

a. orders

b. prompts

c. pretty pleases

3. Which of these is something that AI can't do?

a. find patterns in data

b. write a song

c. understand human emotions

4. Why are there so few laws about AI?

a. because it's still so new

b. because people think computers should be able to do whatever they want

c. because governments are too busy with other things

The answers are on page 24.

GLOSSARY

adapt to change or adjust to new conditions

analyze to examine something in detail in order to find patterns or explain it

app a computer program designed to do a particular job, usually on a smartphone or tablet

chatbot a computer program designed to have conversations with humans

content something such as text or music that is produced for others to read, watch, listen to

data information that is stored or used in a computer, in the form of a series of ones and zeroes

emotion a feeling such as joy, anger, or fear

pattern an arrangement of similarities or trends between different items in a set

program a set of coded instructions for a computer to follow

prosthetic a human-made body part

robot a machine with moving parts that is programmed to do a job

senses physical abilities that let us see, hear, taste, smell, and touch

statistics facts or data, in the form of numbers

therapist a person who is trained to listen to people and help with their problems

Books

Artificial Intelligence. Julie Murray, Abdo Zoom, 2021.

Artificial Intelligence. Dinah Williams, Starry Forest Books, 2021.

Robots and Artificial Intelligence. Nancy Dickmann, Gareth Stevens, 2019.

Websites

www.bbc.co.uk/newsround/49274918

kids.britannica.com/kids/article/artificial-intelligence/390648

kids.nationalgeographic.com/books/article/could-a-robot-become-president

INDEX

Answers: 1. a; 2. b; 3. c; 4. a